Waste and Recycling

Janine Amos

Franklin Watts
London • New York • Sydney • Toronto

Franklin Watts
96 Leonard Street
London EC2A 4RH

Franklin Watts Australia
14 Mars Road
Lane Cove
NSW 2066

UK ISBN: 0 7496 0712 2

Editor: A. Patricia Sechi
Design: Shaun Barlow
Cover Design: K and Co
Artwork: Ian Thompson
Cover Artwork: Hayward Art Group
Picture Research: Ambreen Husain

Educational Advisor: Joy Richardson

A CIP catalogue record for this book is available from the British Library

Printed in Italy
by G. Canale & C. SpA

Contents

What is waste?

Everything that we throw away is waste. So is anything that is left over when factories work to make goods. You can see waste in rubbish bins, in the street and even floating in rivers and seas.

Waste is a problem. We do not want it, yet we are producing more than ever before. Waste in the wrong place spoils our world.

▽ Waste such as wrappers and cans litters this street.

Waste in your home

How much rubbish does your household make? Think of all the wrappers, papers and cans left at the end of the week. How many cleaning products come in bottles which will later be thrown away? And how much leftover food goes into your bin?

▷ Plastic bottles, boxes and cans are soon emptied. But all the containers stay with us.

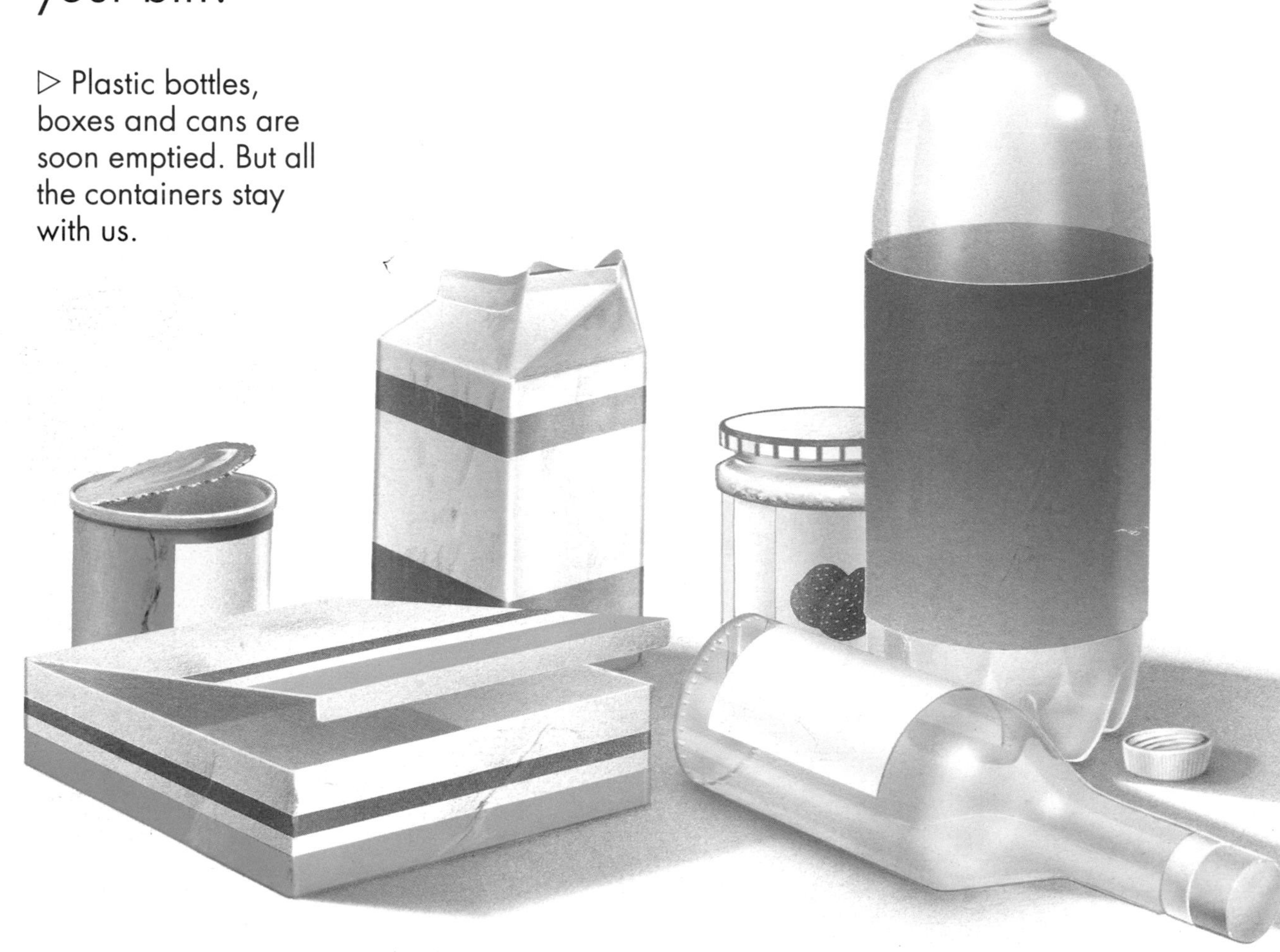

▽ One family has produced all this waste in just one week. Imagine all the rubbish the whole world makes!

△ Every meal leaves a trail of rubbish behind.

The world's waste

In some countries household waste is collected from homes every week. It is taken in lorries to huge dumps called **landfill sites.** All the rubbish from shops, offices, schools and restaurants is collected too. In some areas there is no waste collection. Great piles of rubbish are left to rot in the streets.

▷ In some countries rubbish is thrown on open dumps. It soon begins to smell bad and may become a home for rats.

▽ In many countries, workers take rubbish away in trucks.

▽ At landfill sites the waste is packed into the ground and covered with earth.

More waste

Most of the world's waste comes from farms, mines and factories. Some of this rubbish is dangerous. It may catch fire or be poisonous. Millions of tonnes of dangerous waste are produced every year. Some is burned but this makes the air unhealthy. Some is buried in landfill sites lined with plastic or clay. But poisons can leak out.

▷ This landfill site is lined with clay before it is filled with rubbish.

▽ Poisonous wastes may be stored in drums and buried.

Disaster

The world's waste must be carefully managed. Some kinds of rotting rubbish give off a gas called methane. This gas explodes if it is mixed with air. Houses near dumps have been destroyed by methane explosions.

Slag is waste left over from mining. It is sometimes dumped near mines. But if slag gets really wet, it may slide.

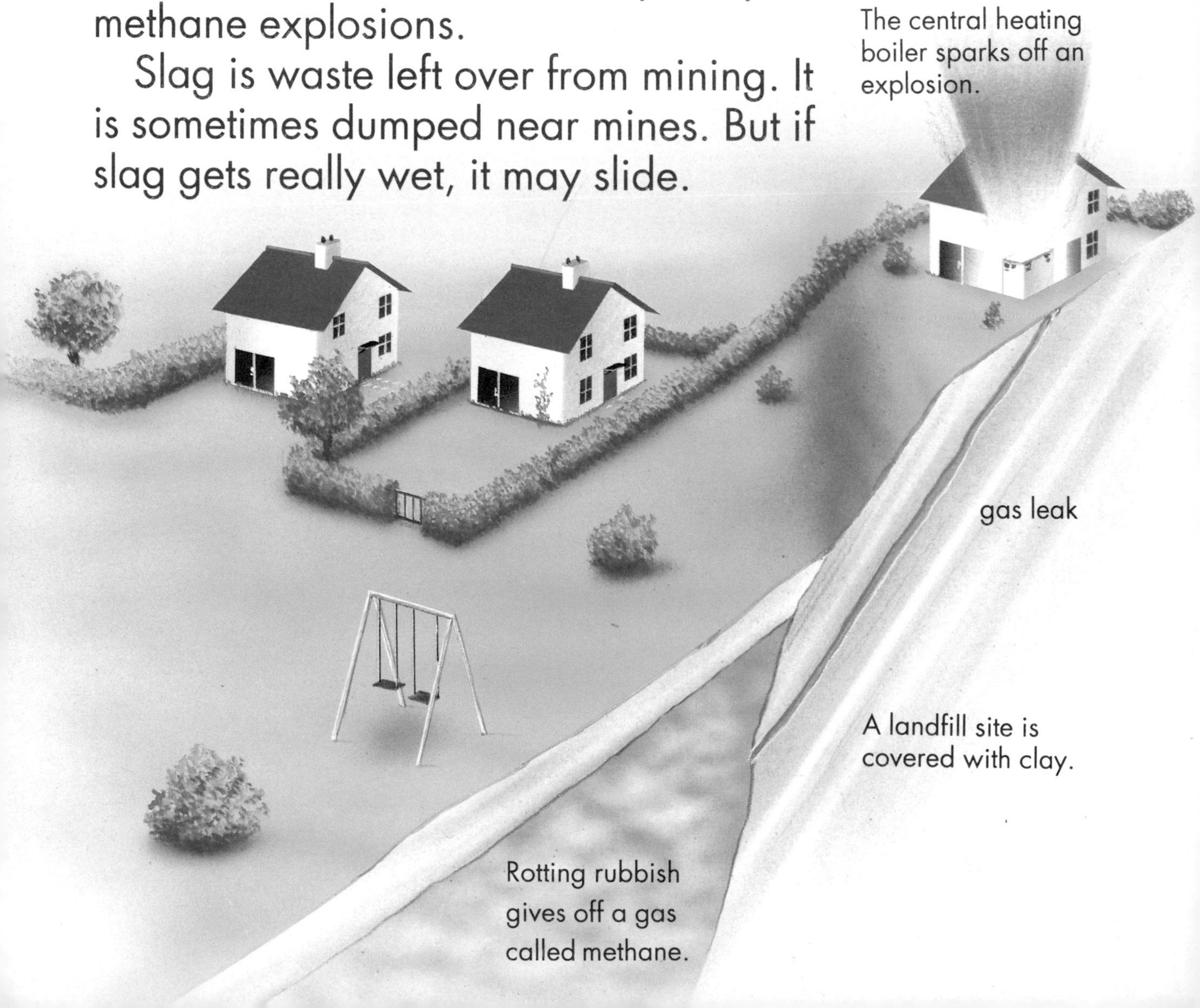

The central heating boiler sparks off an explosion.

gas leak

A landfill site is covered with clay.

Rotting rubbish gives off a gas called methane.

▷ Slag heaps from coal mines spoil the look of the land. They may also be dangerous.

▽ A school and other buildings in Wales were destroyed by slag sliding.

Waste in the water

People pour waste of all kinds into water. Human waste from sinks and toilets is called sewage. Most sewage ends up in rivers, lakes or seas. Waste chemicals from factories are also poured into the oceans. They poison animals. Oil in the oceans kills wildlife too.

▷ Waste is burnt on ships out at sea. This causes pollution.

▽ Huge ships wash out their oily tanks in the ocean.

▽ In some parts of the world sewers are open. This may cause disease and make people ill.

△ Sewage runs straight into some rivers.

Nuclear waste

Nuclear power stations give us power to make electricity. They do not send up dirty smoke into the air. The **spent fuel** and waste they produce are dangerous. The most dangerous waste stays dangerous for thousands of years. The waste must be stored away from people for all this time. Scientists are thinking of new ways to deal with the waste.

▷ Spent fuel is still dangerous and is stored underwater.

▽ Dangerous nuclear waste is kept underground in these modern computer controlled tanks.

P8 1122/716/2
F 23293

Running out

Dumping our rubbish takes up huge areas of land. Land, water, coal, oil and wood are all **natural resources.** Natural resources are valuable because we can never make more of them. Only nature can do that. Every year we use up more and more of our natural resources. But we throw much of them away as waste.

▽ Many cities have no space left for dumping rubbish. It has to be taken to sites out of town.

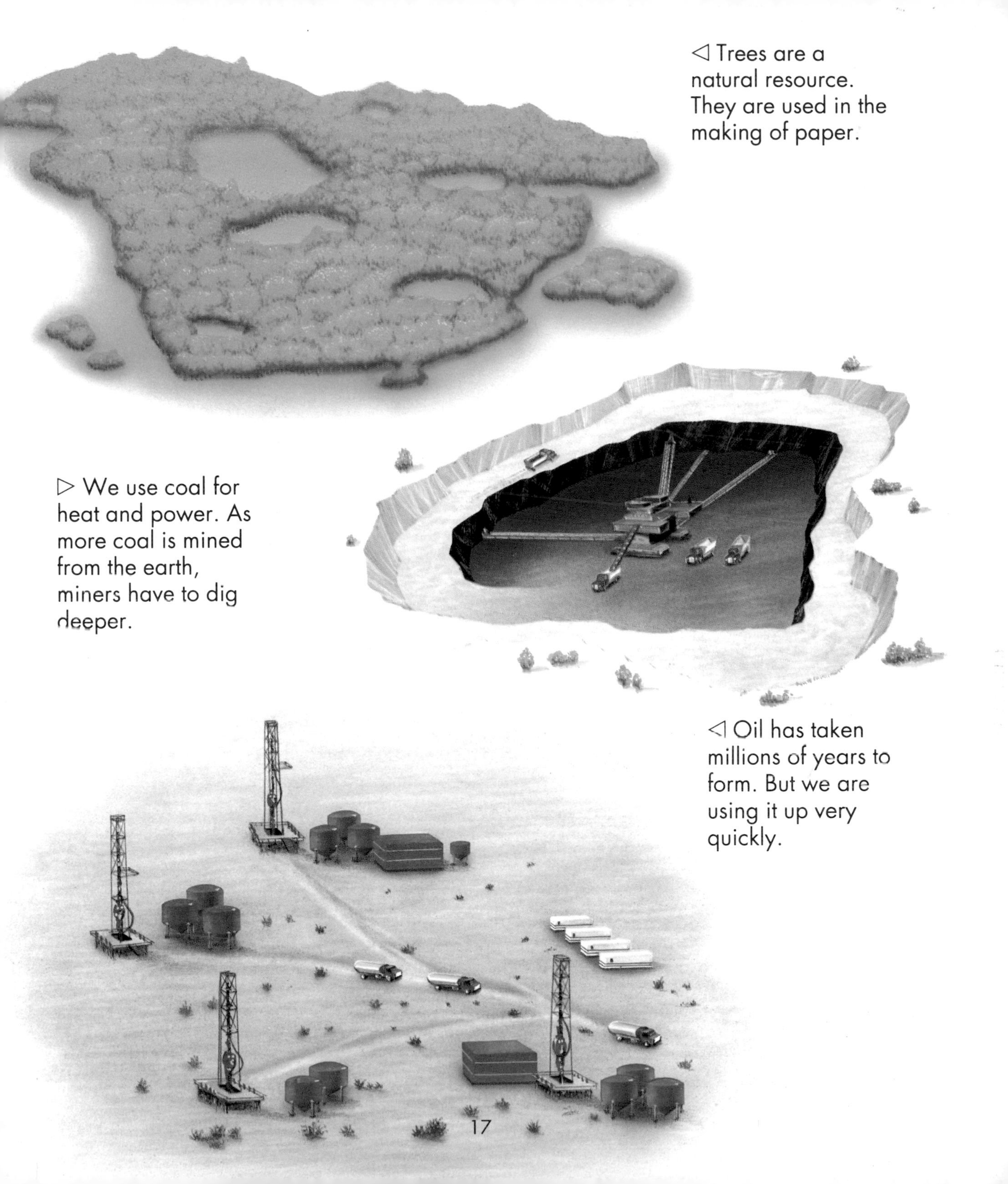

◁ Trees are a natural resource. They are used in the making of paper.

▷ We use coal for heat and power. As more coal is mined from the earth, miners have to dig deeper.

◁ Oil has taken millions of years to form. But we are using it up very quickly.

Packets and packaging

Many of the goods we buy come in decorated packets. They look attractive. Goods for sale must be protected and kept clean. But often there are many more layers of paper and card than we need. All this packaging adds to the world's waste.

▷ Many trees are cut down and pulped to make paper.

◁ Aluminium is used for cans and cooking foil. It is difficult to make and uses lots of energy.

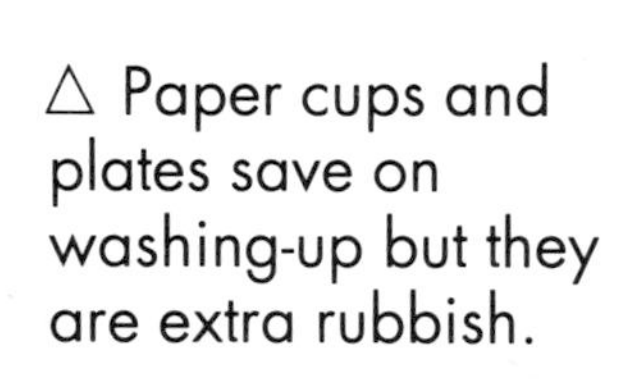

△ Paper cups and plates save on washing-up but they are extra rubbish.

▷ Some goods are packed in fancy wrappers to make them look bigger.

Recycling

Paper, metal and glass products can be used again. First they are sorted. Then modern machinery breaks them down ready to be used to make new products. This process is called **recycling**. In this way waste paper may be used in making wrappers, tissues, newspapers and writing paper.

▽ In some countries, there is one bin for paper and one for glass and one for food scraps.

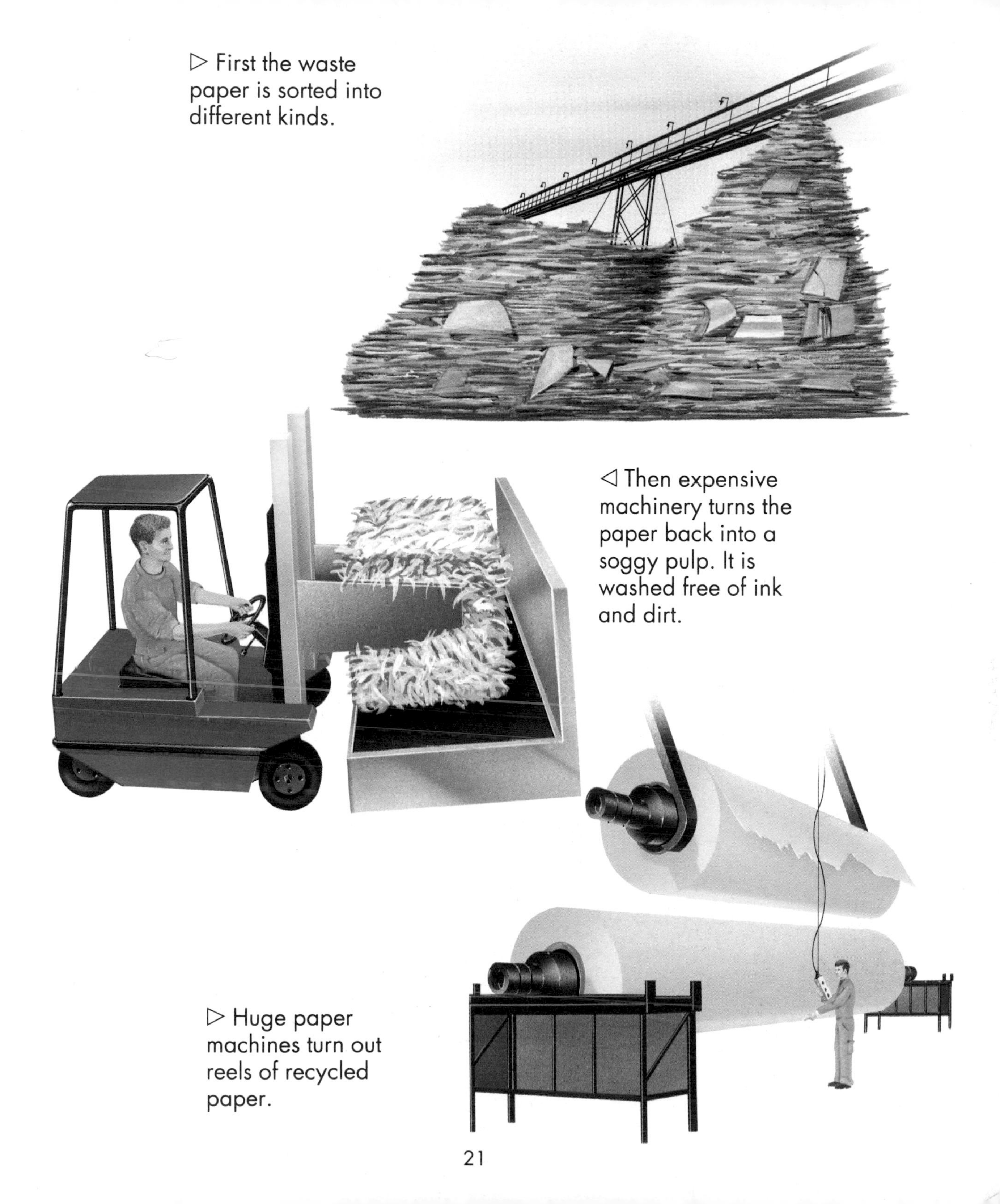

▷ First the waste paper is sorted into different kinds.

◁ Then expensive machinery turns the paper back into a soggy pulp. It is washed free of ink and dirt.

▷ Huge paper machines turn out reels of recycled paper.

Metal and glass

Paper is not the only material which we recycle. Aluminium cans are melted down and used over and over again. Steel is easy to sort out from other rubbish using huge magnets. New steel products are made from the scrap. Broken glass bottles and jars go to make new ones. Recycling saves energy and valuable resources.

▷ Recycling aluminium uses a tiny amount of the energy used to make aluminium in the first place.

▽ Reusable metals are removed from old cars. But there is a lot of waste left over.

▽ Glass bottles from a bottle bank are gathered to be melted down to make new ones.

Wasting water

Clean water is important for drinking and cooking. It is pumped into homes and factories from rivers and **reservoirs**. We use a huge amount of water and we often waste it. Water can be recycled. Dirty water goes along pipes to be cleaned and made ready to use again. As cities grow we need more water.

▷ Some people fetch water by hand from a well. They make careful use of their water.

▽ Dirty water is cleaned in a water treatment centre.

dount waste
whate

Waste and the future

It is becoming more and more important to make less waste. Many factories and mines are already using new machinery designed to cut down on rubbish. Reusing the waste we produce is just as important. All new products must be planned with the waste problem in mind. We all need to think more about what we waste.

▷ Some plastics make rubbish that will not go away. These bags are designed to rot into the soil.

▽ The new machinery in this coal mine allows more coal to be mined with less waste.

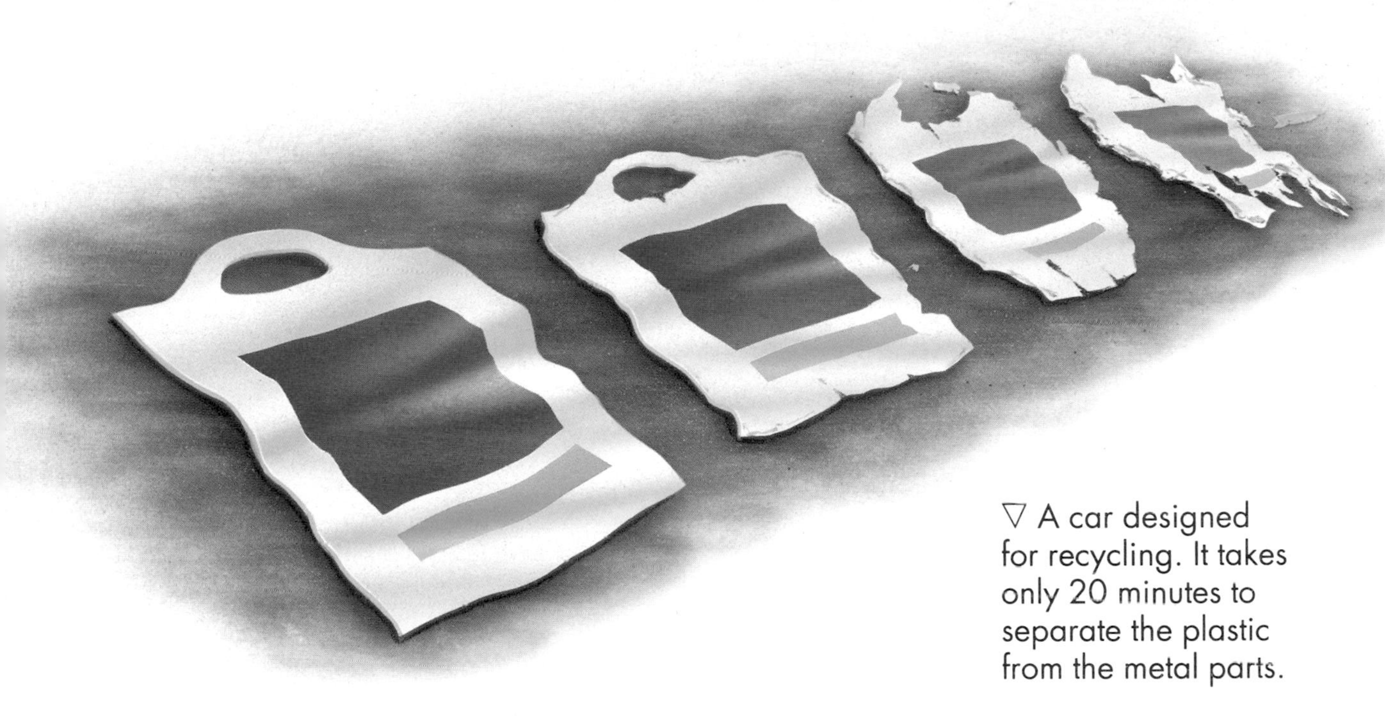

▽ A car designed for recycling. It takes only 20 minutes to separate the plastic from the metal parts.

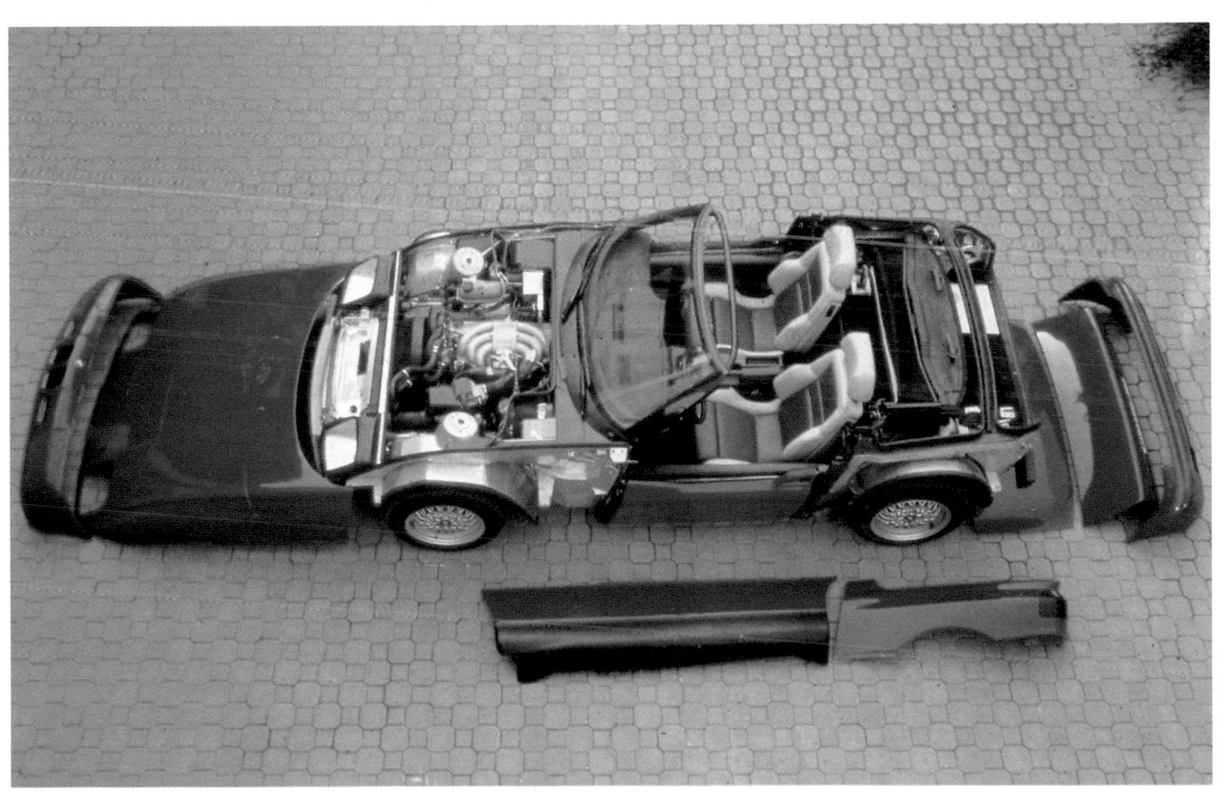

Success stories

People are learning how to manage waste successfully. They are finding ways of putting it to good use. Methane gas from rotting food may be used for heating. In Finland many sawmills now use their own waste wood as fuel. New laws are forcing factories to get rid of their waste safely. Some factories even make money by selling their waste!

▽ These children have made a machine for squashing cans ready for recycling.

◁ In some countries food waste is put into large sealed containers. The food rots and produces methane.

▽ In other countries large amounts of food waste are used to fuel whole factories.

Things to do

Everyone can help to cut down on waste. You can help by:

- Never throwing away anything which can be reused, recycled or repaired.
- Never buying throwaway goods such as paper cups and plates.
- Asking your family to buy recycled tissues, toilet paper, kitchen paper and writing paper.

Useful addresses:

Friends of the Earth
26-28 Underwood Street
London N1 7JQ

Greenpeace
Canonbury Villas
London N1 2PN

Friends of the Earth
Chain Reaction Co-operative
PO Box 530E
Melbourne
Victoria 3001

Glossary

aluminium A light metal

biodegradable Something which is able to rot away.

landfill site An area which has been set aside for waste to be buried in it.

reservoir A lake or large tank used for storing water.

sources of energy The oil, gas and electricity used for lighting and heating.

spent fuel Fuel in a nuclear power station which has been used up after making electricity.

natural resources Natural products of the Earth such as land, water, trees, metals, coals and oil

pollution Dirt or waste which spoils the air, land or water.

recycle To put something through a process where it is broken down so that it may be used again.

Index

Photographic credits: BMW 27; British Nuclear Fuels plc 15; Ecoscene (Cooper) 5, (Hawkes) 9, (Sally Morgan) 19, (Eva Miessler) 20, (Gryniewicz) 23; Mark Edwards/Still Pictures 28; The Environmental Picture Library (Martin Bond) 3, (P. Glendell) 24; Greenpeace (D. Van der Veer) 13; Hutchison Library 7, (Robert Francis) 16; Syndication International Ltd. 11.